English Lavender

For pleasure lavender is surely the
typical English perfume, evocative of
country gardens and days when life
was lived at a slower pace.
The flowers possess a delicious
sweetness when dried and a delightful
perfume is enjoyed in oil of lavender
and lavender water.

English Lavender

'And Lavender, whose spikes of
azure bloom
Shall be erewhile in arid bundles bound,
To lurk amid her labours of the loom,
And crown her kerchiefs clean with
mickle rare perfume'
Shenstone

ENGLISH LAVENDER

Copper Beech Publishing

English Lavender

Published in Great Britain by
Copper Beech Publishing Ltd
©Copper Beech Publishing Ltd 2004

Researched and edited by
Jan Barnes and Beryl Peters

ISBN 1 898617 39 2

A CIP catalogue record for this book is available from The British Library.

Acknowledgements page 61.

Copper Beech Gift Books
Copper Beech Publishing Ltd
P O Box 159 East Grinstead
Sussex England RH19 4FS

English Lavender

England's moist and moderate climate gives a marked superiority to English lavender. The crop is best when a hot summer follows a mild winter.

CALM

Lavender is again being chosen as a perfume and as a fragrance to gently scent the air.

We are rediscovering how scents can affect us and in days gone by, lavender was used in many different ways. As well as the more obvious uses for the sweet smelling herb, lavender's traditional qualities as a soothing restorative are now adding a much needed air of relaxation and calm.

This little book will help uplift the spirits and bestow a feeling of calm to balance the effects of a busy life.

English Lavender

For passion of the heart
The flowers of lavender picked from the
knaps (the blue part and not the husk)
mixed with cinnamon, nutmegs, cloves,
made into powder and given to drink in the
distilled water thereof, doth help the
panting and passion
of the heart.

*English Lavender
Lavender Vera, English
lavender, is a small
growing shrub, with
smooth, narrow and
pointed leaves.
It blooms in summer
and has long spikes of
deep mauve which smell
good. It is ideal for
low hedges.*

IN THE GARDEN

The warm sweet smell of summer
Lavender has been a perennial favourite in English gardens. Nearly every garden has at least one lavender bush. It has kept its popularity from long ago to the present time. Each spray conjures up the warm, sweet smell of summer.

Lavender is in the garden for the bees, for use in the pot pourri that scents the drawing room, and for the washes it makes for the hair and the skin.

The bees and butterflies are attracted to lavender flowers and when the bees are busy it's a sign of harvest time.
It is said that:
'where there's honey there's oil'.

Pruning
Lavender should be pruned after flowering, to ensure a healthy, well-shaped shrub.

Harvest the lavender when ripe
Lavender needs harvesting when ripe and the flowers are fully open all up the spike but before fading has occurred. It is best cut with long stems with a curved-bladed knife to sever a large handful at a time.

*One of the attractive walks in a kitchen
garden is that bordered with flowering
lavender bushes. Traditionally, the
gardener would cut the lavender borders
to release the scent and
the ladies could enjoy the fragrance
of lavender as they walked
round their garden.*

NOTES ON PROPAGATION

Layering
Long stems can be bent down and held in place in the soil with a wooden peg. Once rooted, cut away from the main plant.

Side shoots called 'slips'
Lavender, along with rosemary, sage and thyme is readily increased by pulling off side shoots called 'slips' from the main branches and inserting them firmly in a shady corner, about three inches deep, early in April. Water occasionally in dry weather and let the slips remain untouched until the following April, when they will be sufficiently well rooted to be lifted and replanted.

The flowers we tend with our own hands
have a habit of blooming in our
expectations and filling our hopes with a
sweetness, which, not even the most skilful
hired gardeners have ever taught the
most far-fetched hybrids.
Vernon Lee

ASPIC JELLY

Flavouring ingredient

'And now it may be asked: why is it called aspic? There is upon this point the most curious ignorance, though the explanation lies upon the surface. It means good old English spike, lavender-spike, and spikenard. Lavender spike is to be found in the sauces of Roman cookery; it is mentioned among the pot-herbs used in France five hundred years ago; one of the spikes - the spikenard of Spain - is in English books of the same period mentioned as a flavouring ingredient of Hippocras.'

The Book of the Table 1877

PHYSIC GARDENS

Lavender is a cultivated plant which appears to have first been noticed in England about 1568.

Herbalists

The physic gardens were the first botanical gardens for city dwellers to see herbs and plants growing. Herbalists would teach people the properties of each of the plants. These public gardens were invaluable for students of medicine and apothecaries began to increase.

SUNDIALS IN THE HERB GARDEN

Sundials were placed in the herb gardens. Those who had the time to sit at rest by the sundial could breathe in the sweet scent of lavender and the other herbs to enhance their enjoyment of the garden. Charles Lamb said 'Sundials are so ancient, Adam must have had one in paradise.'

Warm sun brings out the essential oil content in many plants but too great a heat and drought can cause poor quality.

DRYING THE FLOWERS

Sachets and bags

Dry lavender carefully and fill sachets and bags for all the cupboards and drawers in the house.

The correct way

Harvesting and drying herbs in the correct way is important. Lavender must be gathered before the flower turns brown, for then the aromatic oil is at its best.

A good way of making trays

For sweet bags, perfumes or pot pourri, spread the lavender on trays to dry. A good way of making trays is to stretch a piece of canvas or muslin across a child's hoop. Suspend this in a warm room where there is plenty of air.

IN THE DRAWING ROOM

The jar
There are many uses for dried lavender in the drawing room. The jar of pot-pourri needs replenishing with lavender each year.

Linen
Always remember, some lavender must be set apart for the linen-press.

Lavender oil with beeswax rubbed into wood will prevent infestation. Also, lavender is perfect for the indoor winter flower arrangement.

LAVENDO
FURNITURE
POLISH

Lavender keeps its scent, like rosemary.
Without it, pot pourri is not complete.

A Simple Pot Pourri

This is a way of making a simple form of pot pourri. Chop up some carnations, rose, orange flower blooms, thyme, rosemary, marjoram, myrtle, and lavender. Add a dozen cloves and two ounces of pounded orris root. Then press the chopped ingredients into a deep round bowl with a tight-fitting lid, placing them between layers of pounded bay salt. Keep the pourri closely covered for several weeks, then stir up.

The flower petals should be absolutely dry before being used.

Pot Pourri with Damask Rose

Gather together Damask rose petals and buds, the red parts of clove, gillivers, lavender blossoms, orange flowers, rosemary flowers, jasmine, sweetbriar and wild thyme.

Directions bid us lay all these sweet things in the sun for a time, then strew them in layers in a big china bowl with a sprinkling of bay salt between each layer.

Afterwards, sprinkle over sliced orris root, benjamin and storax.

As a final touch, stick three seville oranges with cloves and then the pot pourri will be complete

IN THE BOUDOIR

Refined women

In the boudoir, many refined women have their drawers and shelves lined with a thin satin quilt of a delicate colour wadded with scented bags.

Sweet bags are similar to lavender bags. They can be filled with a variety of ingredients and while lavender bags are for scenting drawers and linen cupboards, sweet bags can also be used under the pillows and hanging on the backs of chairs, where the warmth and pressure of heads releases the perfume.

Sweet Scented Bags

Take two handfuls each of dried lavender flowers and dried rose petals, add two ounces orris root powder, a little common salt, two ounces coriander seeds, two teaspoons cinnamon, one teaspoon ground cloves and one handful of dried orange blossom. Mix together and fill bags.

Ladies have for years added lavender bags to their delicate linens which emit a delicious, relaxing odour.

A Very Pleasant Perfume

Take one drachm of musk, four cloves, four ounces lavender seeds, one drachm ambergris. Heat your pestle and mortar and rub musk, cloves and lavender seed with a lump of loaf sugar and a wine glass of rosewater. Take a handful of powder, mix it well together, then sift it through a sieve. Add two or three more handfuls of powder until perfume is brought to the proper strength.

Take little white leather bags, the seams well sewed up with cat gut or waxed thread, and keep perfume therein.

Lavender Bags

Lavender bags may be made from organdie or butter muslin. This is cut to the size required and the sides run up in the ordinary way. A bunch of lavender is inserted, heads downward, and the opening made fast with cotton firmly stitched then tied with narrow, mauve ribbon.

This method has the advantage of retaining the perfume longer than if the stalks were removed.

A LADY'S PARAPHERNALIA

A lady's paraphernalia and smalls
Gloves, lace handkerchiefs and stockings can also benefit from the scent of this sweet smelling herb. Ladies have for years found lavender bags the natural way to scent their delicate linens.

Easily prepared sachets
Sachets are easily prepared, too. You can simply sprinkle, more or less abundantly, square pieces of cotton wool with the perfumed powder you like best. Sew these up and trim as prettily as possible. A lavender coloured ribbon would be the perfect choice.

Bonnet-boxes should always contain a delicately scented sachet.

Lavender Bags

Pick lavender heads before they are in full bloom. Gather when the dew has gone and before the hot sun has drawn the scent from the blossoms. Spread in a cool, airy place to dry, *never* in direct sunlight because sunrays draw out much of the valuable aroma even when picked. When dry, rub the tiny flowers from the stalks and fill the bags – or store in an airtight jar. To make the bags, use lavender coloured muslin or organza. Draw the bag in near the top with a lavender ribbon tied in a bow. A little embroidery enhances appearance.

IN THE BATHROOM

Scent the hot water
The Romans are said to have used lavender to scent the hot water in public baths. The name of lavender is derived from the Latin word *lavendum* meaning 'fit for washing'.

An aromatic bath
Boil for five minutes in water one, or all, of the following plants – bay leaves, thyme, rosemary, marjoram, lavender, wormwood, balm and others. Strain and add a little brandy. Shake some into the bath.

'Then let us meet here, for here are fresh sheets that smell of lavender, and I am sure we cannot expect better meat, or better usage in any place.'
Izaak Walton, Compleat Angler, 1653

Lavender Water

Put a handful of lavender flowers into a kettle with some water. Bring it to the boil and keep it gently boiling. Fix a piece of rubber tubing about two to three feet long on the spout of the kettle. Arrange the tubing so that it will dip in and out of a bowl of cold water. The steam will condense in the tube, and the liquid so obtained will be distinctly perfumed with lavender. Collect the lavender water in a bottle as it runs out of the tube.

During the seventeenth century, lavender oil was added when making soap to overcome the rancid smell of it.

IN THE SPARE ROOM

The sheets should always smell of lavender
It is hard to imagine a house, however humble, without its welcoming spare room. And how fresh and clean this spare room should look. Any stain or bad odour is inexcusable amid fragrant surroundings! The sheets should always smell of lavender and there should be a jar of dried lavender as a comforting welcome.

WELCOME GIFTS

Pave the way for a visit
Scent bags are welcome gifts for friends, and the poor always welcome a lavender bag with a thoughtful message attached. In hospitals and in district visiting these small gifts have a distinct use, for they often pave the way for a visit that otherwise would not have been made.

A lavender bundle thrown on the fire will sweetly perfume the whole room.

LAVENDER STALKS AS INCENSE

Never throw away lavender stalks. Dry them and then put them in a jar filled with a saturated solution of saltpetre. Leave them for seven days, then dry slowly and tie in bundles. One stalk lighted will smoulder slowly and throw off its scent.

Queen Victoria was a lover of lavender and Miss Sarah Sprules was 'Purveyor of Lavender Essence to the Queen'.

KEEP OFF THE CLOTHES MOTH

Remember that plenty of lavender in an unused room will keep off the clothes moth, the creature so poetically called 'Silver Lady' although we all know her behaviour among our possessions is far from that of a lady.

*In Elizabethan times, a mixture of fresh
herbs, including lavender, was strewn on
top of the rushes on the floor. When they
were walked on, the air would be
freshened and sweetened by the
strongly aromatic herbs.*

HOW TO MAKE A LAVENDER BOTTLE

Cut off the heads of sweet smelling lavender sticks and place them in a small piece of cotton wool about four inches long. Roll up the cotton wool and tie it tightly round with a piece of cotton, keeping the top and bottom tighter than the centre. This is the foundation for the bottle.

Now take an uneven number of the lavender sticks, nine, eleven or thirteen and cut them into exactly the same length. Place the ends of the sticks round the rolled-up piece of cotton wool, about half an inch down, and tie them very firmly round with cotton. Then bend the long ends which are left back over the whole length of the cotton wool.

Tie firmly at the end of this, keeping them about the same distance apart round the centre.

Next, take a piece of narrow ribbon of any pretty colour, about two yards long, and with the help of a bodkin, thread it in and out of the sticks until the whole of the cotton wool is covered. If you use two colours the effect can be stunning. Blue and violet are very pretty. Thread each colour on a separate bodkin and then take alternate bodkins for each row.

Finish off the ribbon at both ends firmly with a needle and cotton, and cover it with a little ribbon bow of the same colour. Then tie a piece of ribbon round the ends of the sticks, about a couple of inches from the bottom.

'And still she slept an azure-lidded sleep,
In blanched linen, smooth, and lavender'd.
Keats – The Eve of St. Agnes

Stays too tight?

Lavender was also useful when ladies, with their stays too tight, took a fit of fainting. They would cool their forehead with the lavender water to refresh themselves to continue their journey.

THE LADY'S COMPANION

On her travels

World wide travel became popular for ladies in Victorian times and these ladies found lavender an ideal companion, as along with the excitement of newly discovered places, lavender has always acted as a nostalgic reminder of home.

The gentle scent of lavender brings to mind a picture of long, summer evenings at home when lavender would perfume the gardens and the country lanes.

Strange bed

A lavender sachet could be tucked into a strange bed to give the young ladies the familiarity they often yearned for at night time.

A lavender poultice

Adventurers with grazed knees and hands after a stumble could apply a soothing cool lavender poultice to the hurt. It was also a common practice to enclose a lavender poultice under the bandage when supporting a sprained ankle or wrist. A drop of lavender oil would be placed on the temple to reduce the anxiety caused by the injury and to give quick relief to the ensuing headache and swooning effect. Ladies of 'a certain age' were well used to combatting the vapours by a dab of their lavender water on the temple.

The over zealous hiker

For the over zealous hiker, lavender water could be added to a bowl of tepid water and used as a foot soak.

The experienced traveller would know how useful lavender oil could be.

When packing the trunk

A good lady's maid would always remember to include a lavender bag when packing the trunk so that the linen and clothes would also be sweetly scented.

Too much sun

For those who had indulged in too much sun and needed some cool, soothing lotion, lavender and camomile oils could be mixed. Together, they brought great relief and eased the tension of the skin. If any parts of the skin were blistered by the sun, the oils would encourage the cells to heal - much as they do when applied to scalds, small cuts and bruises. Lavender helped to take away anxiety and when those travellers, tired and exhausted from their efforts, lay down for a rest before dressing for dinner, the soporific effects of lavender soon lulled them into state of drowsiness and sleep.

Lavender became known as the herb of
'cleanliness and calm'.

REMEDIES, BALMS & CURATIVES

Restore the spirits
Restore the spirits during the long winter.
A lavender bag will inspire memories of
long summer days, and gardens in flower.

Fainting
Oil of lavender is useful for overcoming
faintness.

The Romans
There is some doubt as to whether the
Romans first brought lavender to Britain,
but we know that wherever they settled
they planted herbs and used them for
soothing and healing. They also rubbed
themselves with lavender oil to keep
insects away as they were marching and
used it for massaging aching limbs.

REMEDIES, BALMS & CURATIVES

Snakebite

If the skin is stung or bit with a snake, take lavender water and warm it; wash the affected skin morning and evening.

For the eyes

Centuries ago lavender was used to 'comforte' and clean the eyes.

Colic

An infusion of lavender flowers can be used in cases of colic and dyspepsia, particularly where these are caused by depression.

Sleep

A drop of lavender on the pillow is said to induce restful sleep.

REMEDIES, BALMS & CURATIVES

Anti-depressant

Lavender is regarded by herbalists as an anti-depressant.

Antiseptic

Lavender oil has healing and antiseptic qualities. It is an essential oil that can be used neat on the skin.

Insect repellent

Lavender flowers and leaves are considered insect repellents.

Eczema

Total abstinence from alcohol should be enjoined in all such skin disorders; but it will often serve a useful purpose to take a dessertspoonful of pure lavender water two or three times a day.

*Quintessence of lavender possesses great
medicinal virtues and is particularly
serviceable in vaporish
and hysteric disorders.*

A Closet for Ladies 1693

THE LADY OF THE MANOR

Storing the herbs

Still rooms were managed by the lady of the manor. She saw to the correct storing of the herbs for the year.

She mixed and prepared herbs and spices and would often combine them in such a way as to command a certain fame in the surrounding villages.

She would use the herbs medicinally to see to the welfare of the families living and working on her husband's land. She would also suggest their use in cooking. She would oversee the giving of whatever lavender and herbs could be spared to those unfortunates who had no lavender of their own, for to share the harvest of the garden was a good custom that went back to the very earliest of days.

"Grate" Idea for the Summer Months.

Lavender leaves and flowers can be added to fruit salads, summer punches and wine cups.

COOKING WITH LAVENDER
RECIPES AND REFLECTIONS

'Excellent herbes'

Queen Henrietta Maria, wife of Charles Stuart and Charles II's mother had 'very great and large borders of rosemary and rue and white lavender, and great variety of excellent herbes' in her garden at Wimbledon Manor. She liked lavender conserve and lavender wine.

Preserving in vinegar

Herbs can be preserved in vinegar or oil. Wine or cider vinegar is best. Fill a glass jar with herbs or a herb of your choice and fill with vinegar. Stand in the sun for about a week to transfer the aroma of the herb to the vinegar. Strain, bottle and label. Put a small sprig of the herb in the bottle to add interest.

Preserving in oil

Herbs can be preserved in olive oil easily. Fill a glass jar with herbs, cover in oil, stir to get out any air and seal. The oil will soon become aromatic.

A *favourite* of Queen Elizabeth I

Lavender flowers, steeped in sugar and used in conserves to be served at table, were a favourite of Queen Elizabeth I. The conserve had internally warming and comforting properties. Conserve was made by stripping lavender flowers from their stalks, pounding them with three times their weight in sugar and pressing the mixture into pots.

Lavender jelly is delightfully aromatic.

Lavender sugar

Lavender sugar can be used to make ice cream or to flavour puddings or biscuits. Use sparingly. Use freshly picked lavender when it is just coming in to flower, before full bloom. Spread it out to dry, remove the stalks. Blend with four times their weight in sugar. Then pour into air-tight jars. Rose petals or mint can also be used to flavour sugar by the same method.

To candy lavender flowers

Take your lavender flowers ready picked and weigh them. To every ounce of flowers you must take two ounces of hard sugar and one ounce of sugar candy. Boil well. Then put in your lavender flowers when the sugar is almost cold. Stir them together till they be enough, then take them out and put in a box.

A Closet for Ladies 1693

Lavender shortbread

520 gm of plain flour (1lb 2 oz)
520 gm of soft margarine (1lb 2 oz)
240 gm caster sugar (8 oz)
240 gm ground rice (8 oz)
2 level teaspoons dried lavender flowers.
Makes approximately thirty pieces.

Method
Mix all ingredients together, beating well until everything is incorporated and leaves the sides of the bowl.
Divide the mixture between five greased and floured seven inch sandwich tins. Bake in a moderate oven for about 35 minutes. Cut each round into eight equal portions as soon as possible after leaving the oven. Leave to cool completely in the tins. Sprinkle with caster sugar.
Recipe reproduced courtesy of Norfolk Lavender.

Lavender wine

To make lavender wine – two ounces of lavender flowers and three ounces of candy into a bottle of sack 'and shake it oft, then run it through.'

Lavender vinegar

This gives a subtle flavour to salads. Steep lavender heads in one pint of white vinegar in a glass container on a sunny shelf for two weeks. Strain and use.

HISTORICAL NOTES

Lavender is said to be the 'spikenard' mentioned by St. Mark.

In Tudor times

In Tudor times, its strong fragrance deterred vermin in the bed and warned off moths. Air fresheners were made simply by hanging bunches of lavender in bags around the rooms.

Healing herbs

Monks healed with herbs and used those grown in the 'physick' gardens, in their hospitals. Lavender was one of the most popular because of its scent and easy growth. The Melton Priory grew lavender as early as 1301 as records show that 'Jon the Gardener' mentions it in his treatise in verse called 'A Feate of Gardening'.

Lavender sellers

Lavender was used extensively during the Great Plague of London, not only as a disinfectant, but it also staved off the stench. There were many lavender sellers because some of the most famous lavender fields were situated just outside London. The lavender girls would sell their lavender along the busy London streets.

During the plague of London in 1665, lavender was burnt in bundles in churches, and halls where crowds gathered. It was thought that lavender scented kerchiefs, held up against the nose would help to prevent people succumbing to the plague.

Lavender was also an ingredient in *'Four Thieves'* which was a vinegar valued as a protection against the plague. The name is derived from a group of four robbers who confessed to having stolen from plague victims. They protected themselves from infection by wearing masks of lavender vinegar.

The antiseptic qualities of lavender were put to the test during the First World War when medical supplies were short. Country wide collections of lavender were used to make an antiseptic oil. The oil was used with sphagnum moss as a dressing.

'Nosegay Fan' Barton of Drury Lane
The normal meagre living eaked out by the women and children sellers increased as the asking price of lavender soared. One famous young lady who sold lavender and other herbs was called 'Nosegay Fan' Barton of Drury Lane. She would call out one of the popular street cries such as:

'Won't you buy my sweet lavender,
Sixteen bunches for only one penny?
You'll buy it once, you'll buy it twice,
It makes your clothes smell very nice.'

or

'Lavender, sweet lavender,
Who'll buy my sweet lavender?
Two bunches a penny, Sweet lavender.'

' - Here's flowers for you;
Hot lavender, mints, savory, marjoram;
The marigold, that goes to bed
with the sun,
And with him rises weeping...'
The Winter's Tale (Act IV scene iii)
William Shakespeare

Acknowledgements:

Lavender shortbread recipe reproduced
with permission from:
Norfolk Lavender Ltd
'England's Premier Lavender Farm'.

Bunches of lavender cover picture
used with permission from
Purple Haze Lavender Farm
www.purplehazelavender.com

English Lavender

THE ETIQUETTE COLLECTION *Collect the set!*
ETIQUETTE FOR COFFEE LOVERS
Fresh coffee - the best welcome in the world!
Enjoy the story of coffee drinking, etiquette
and recipes.

ETIQUETTE FOR CHOCOLATE LOVERS
Temptation through the years.
A special treat for all chocolate lovers.

THE ETIQUETTE OF NAMING THE BABY
'A good name keeps its lustre in the dark.'
Old English Proverb

THE ETIQUETTE OF AN ENGLISH TEA
How to serve a perfect English afternoon tea;
traditions, recipes and how to read your fortune in
the tea-leaves afterwards.

THE ETIQUETTE OF ENGLISH PUDDINGS
Traditional recipes for good old-fashioned
puddings together with etiquette notes for serving.

ETIQUETTE FOR GENTLEMEN
*'If you have occasion to use your handkerchief
do so as noiselessly as possible.'*

www.copperbeechpublishing.co.uk